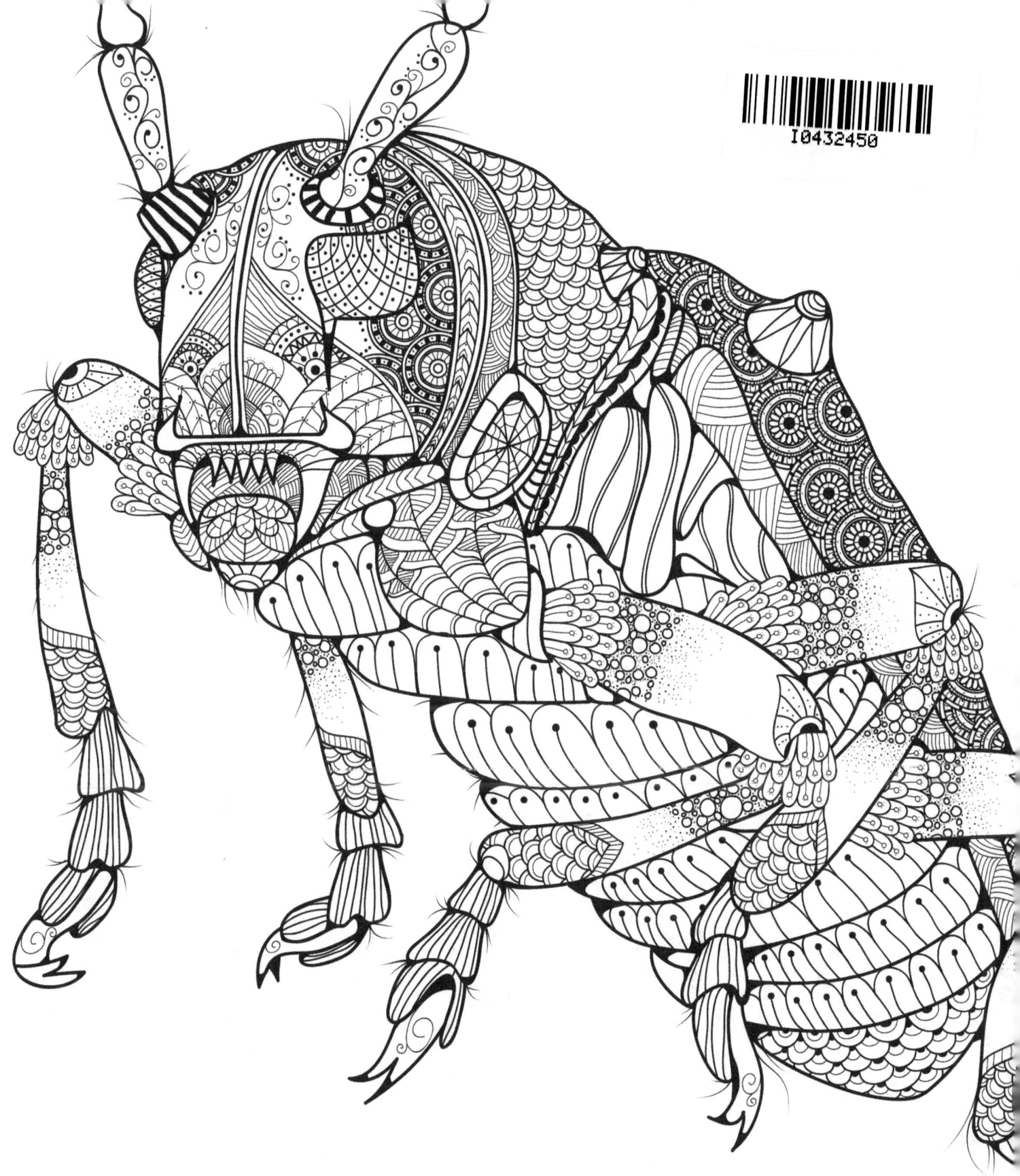

Special Thanks

Wendy Peel

Exclusive Customer
Who Vote This Book Cover Design

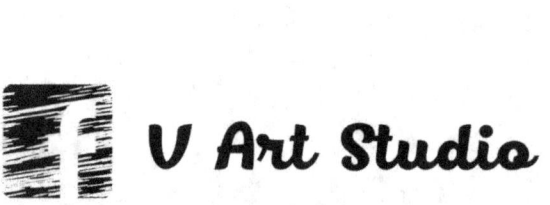 

Sign Up For Get Free Coloring Pages
&
New Release Review Copies

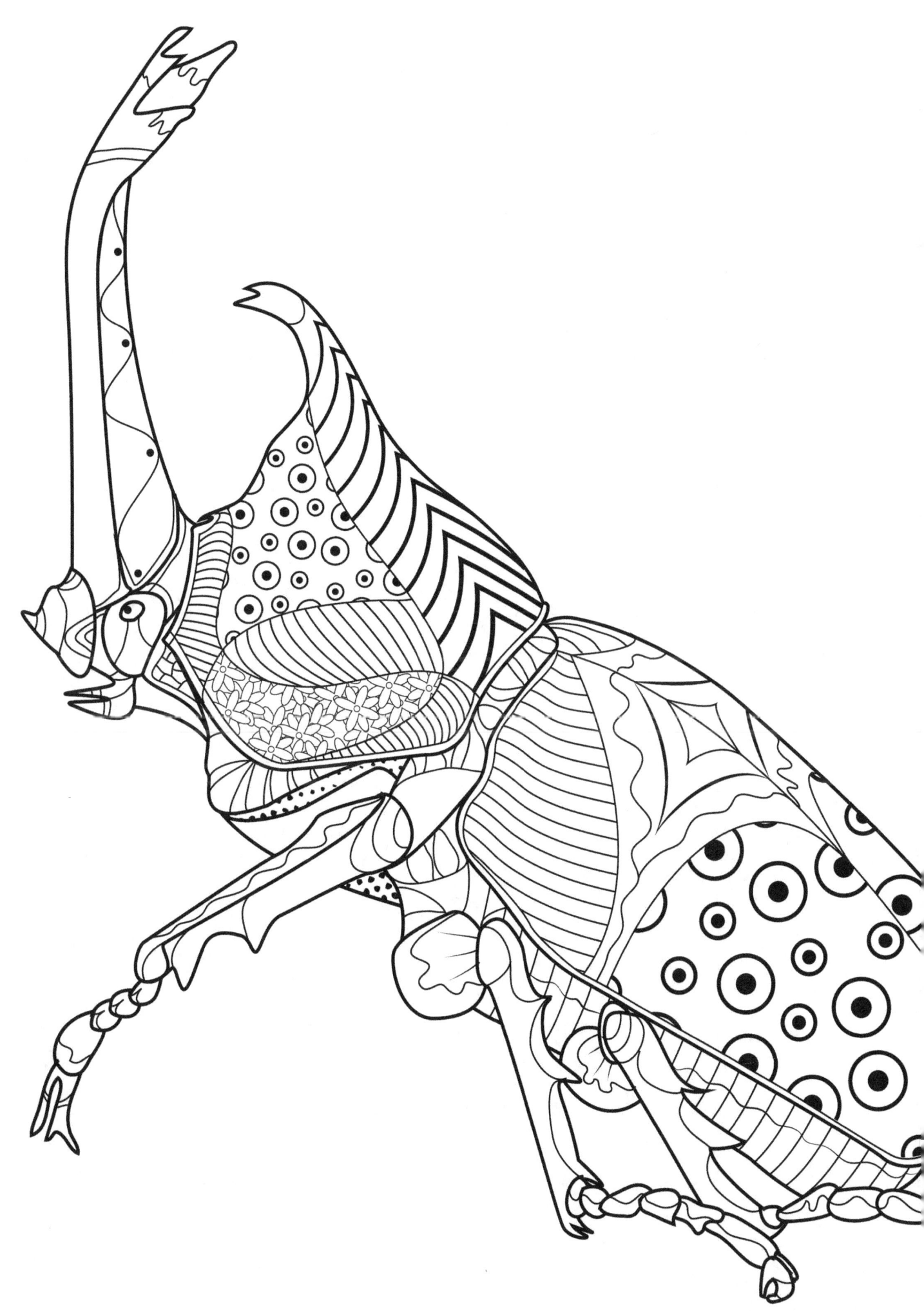

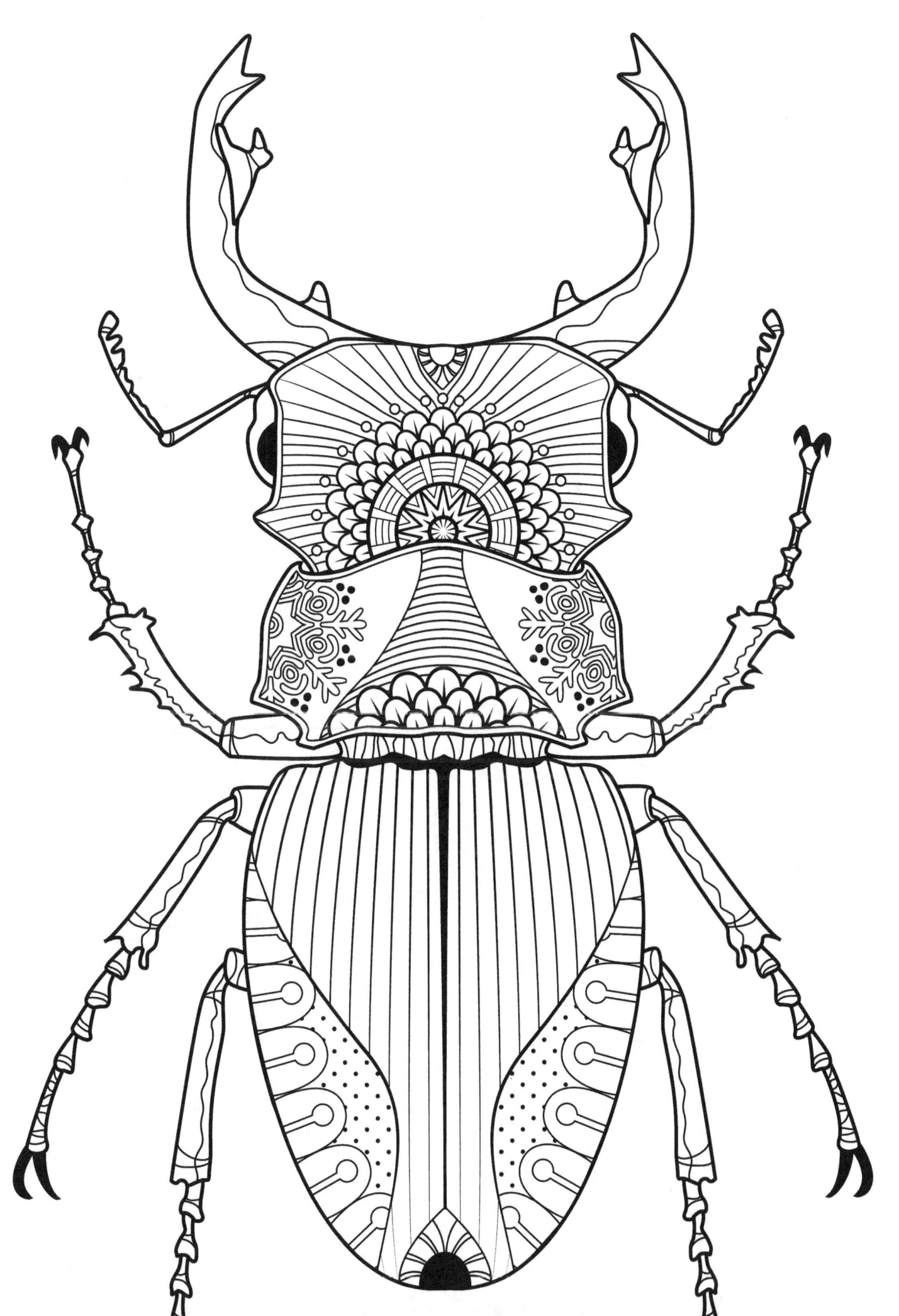

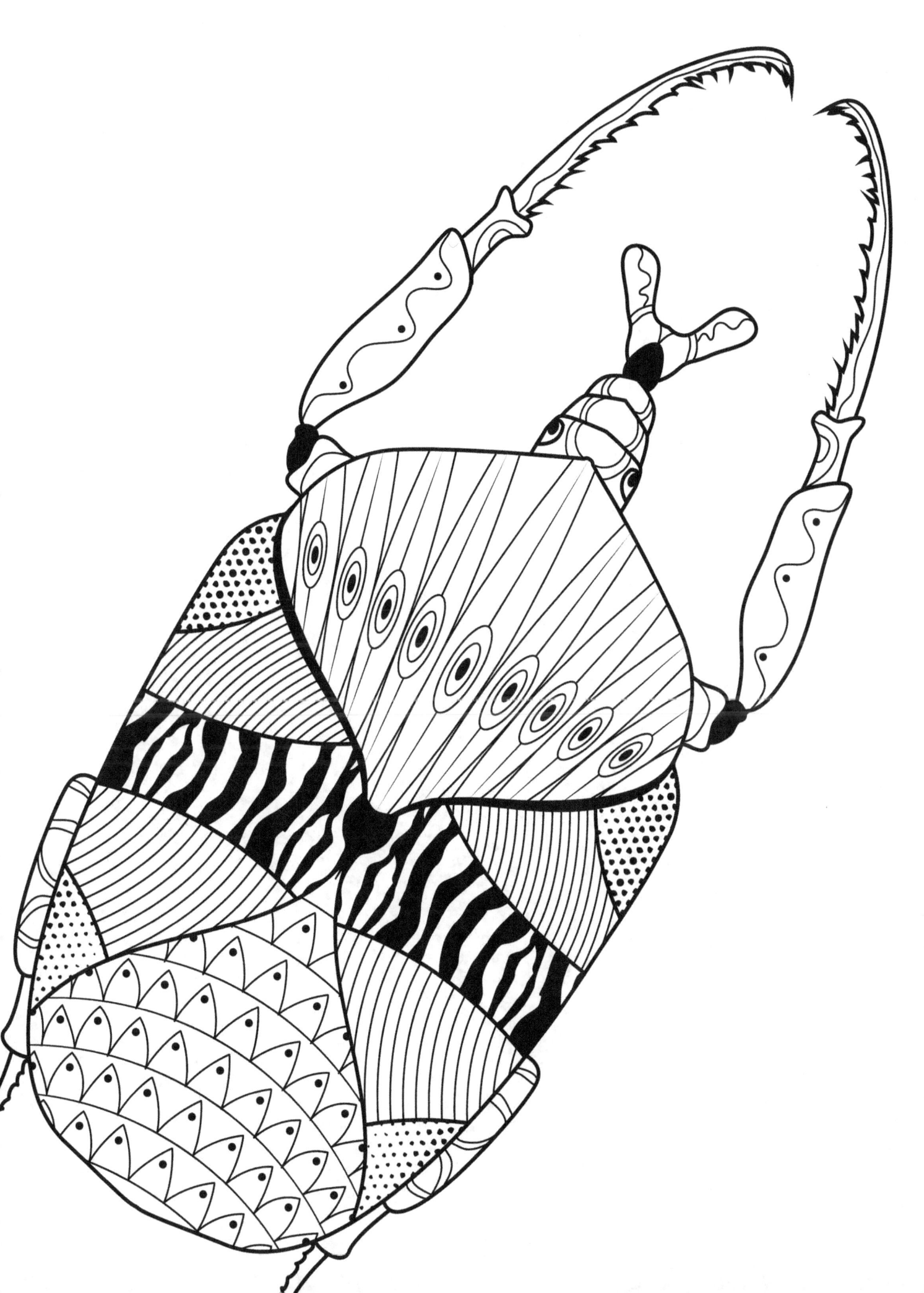

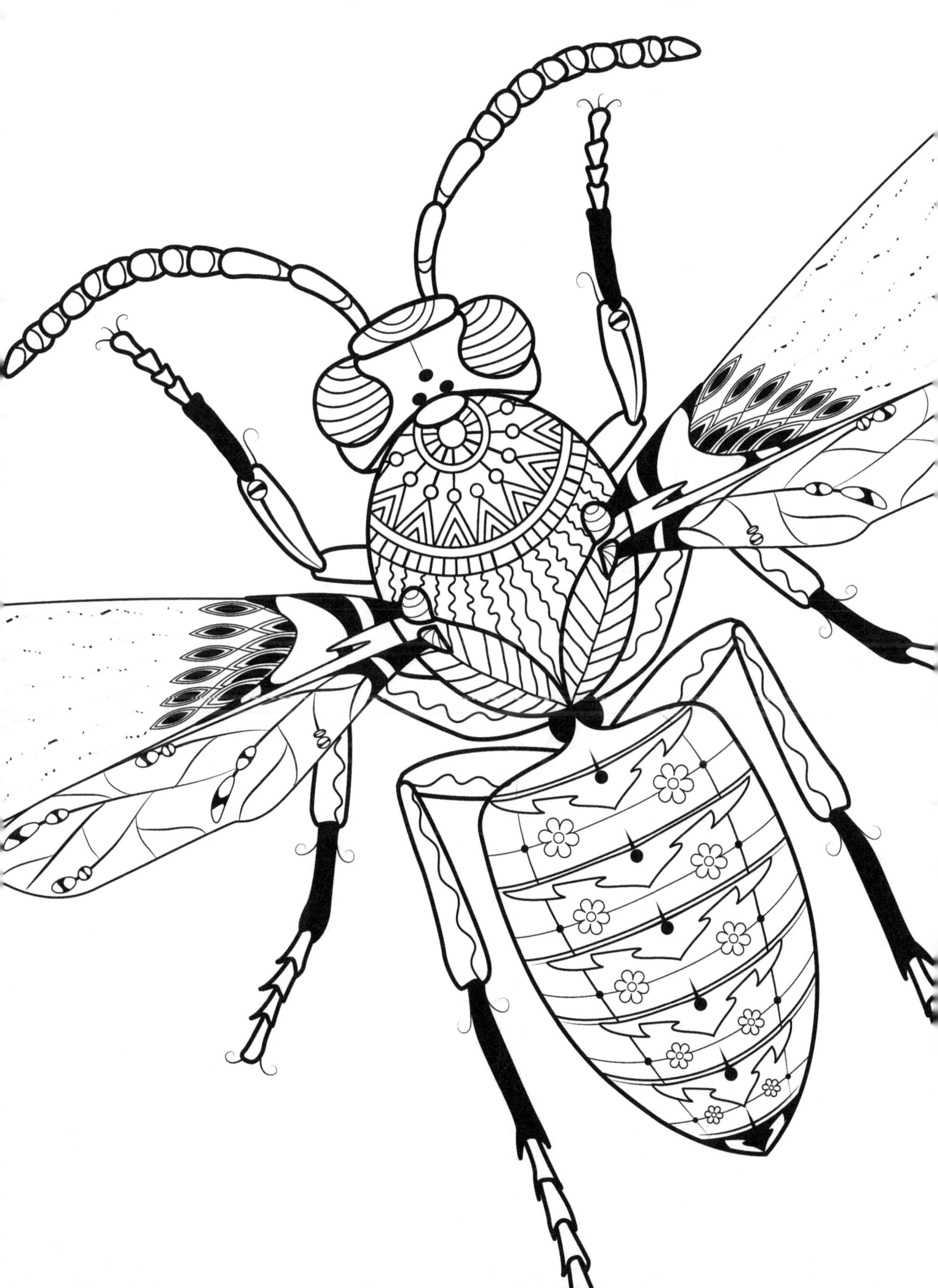

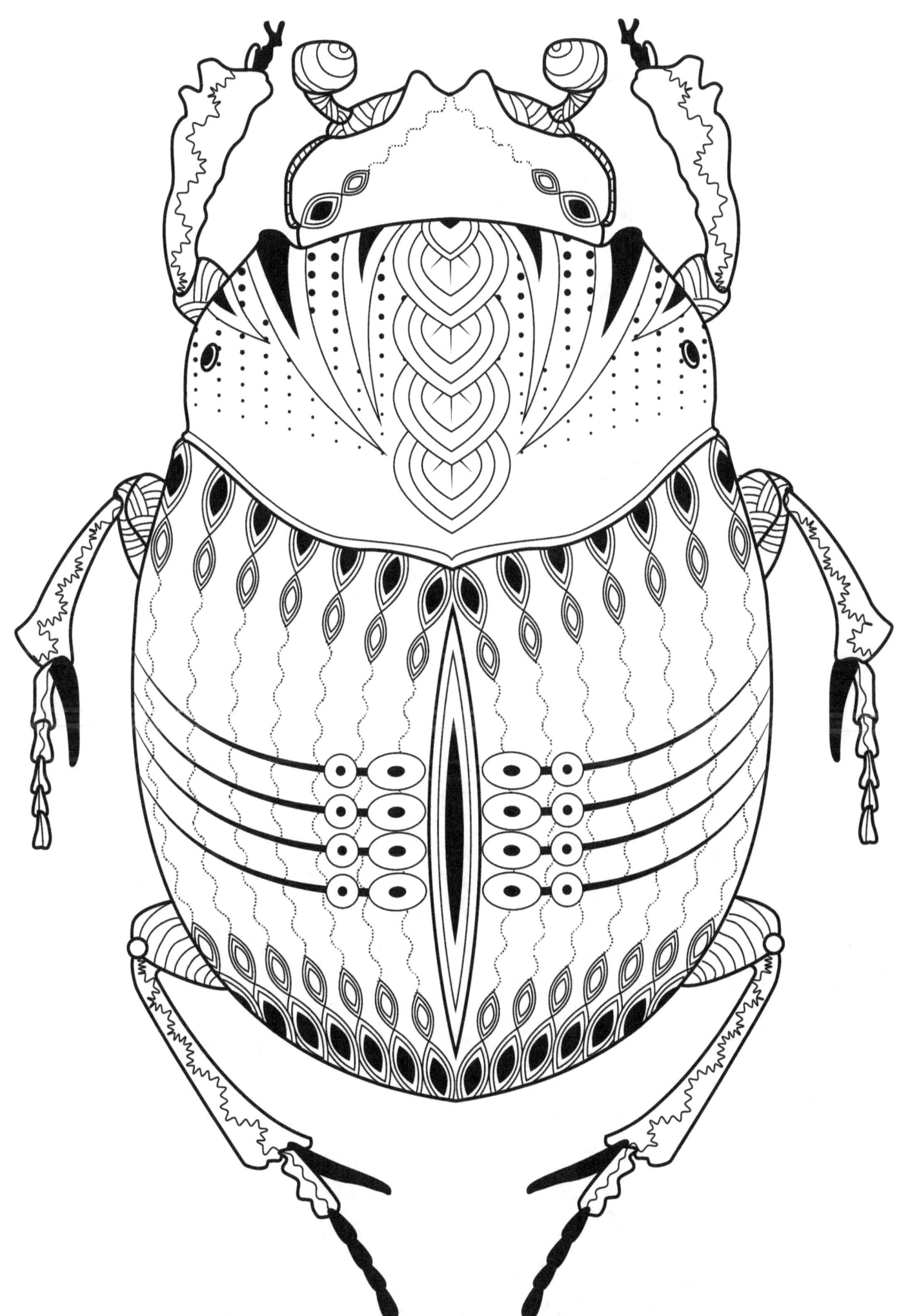

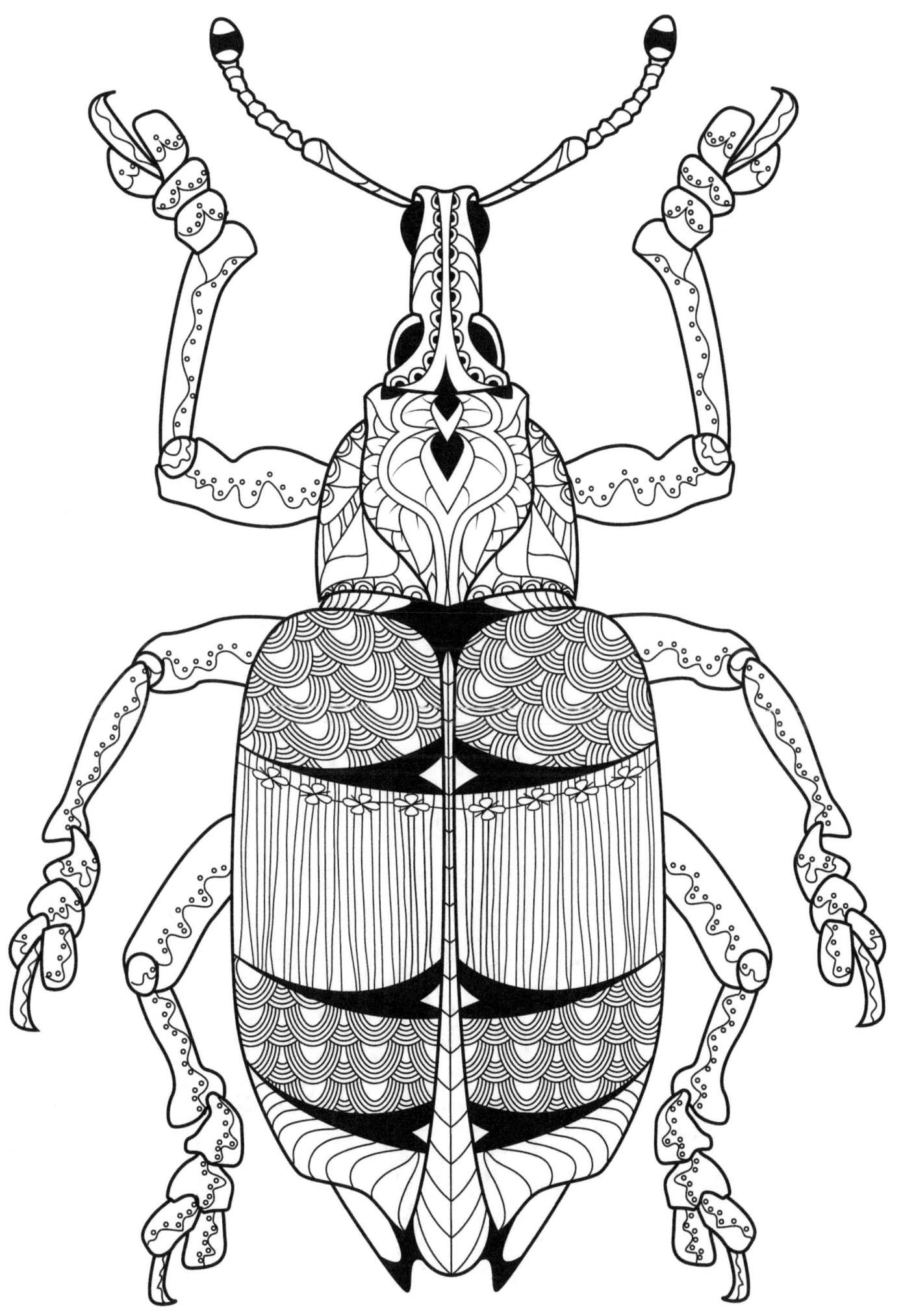

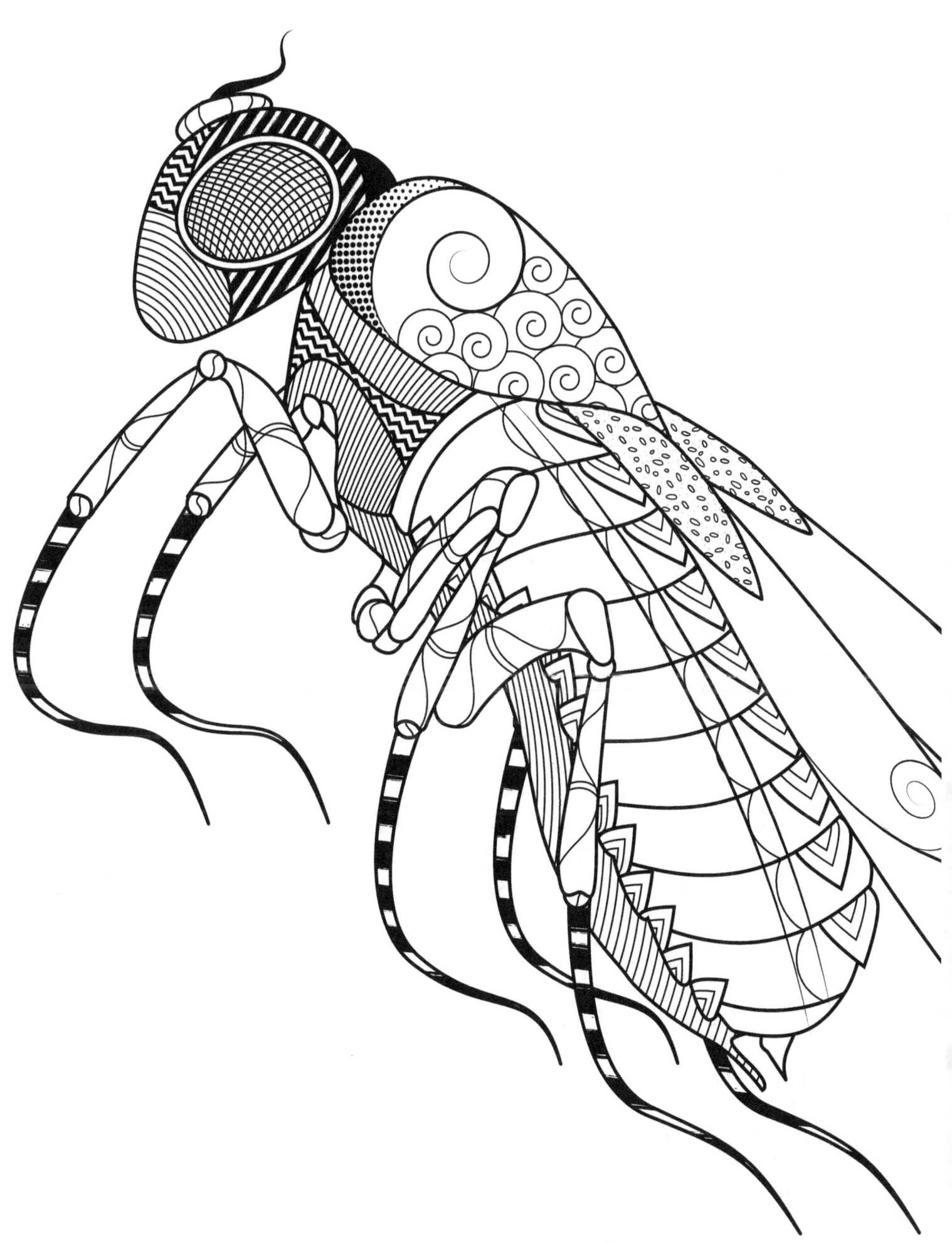

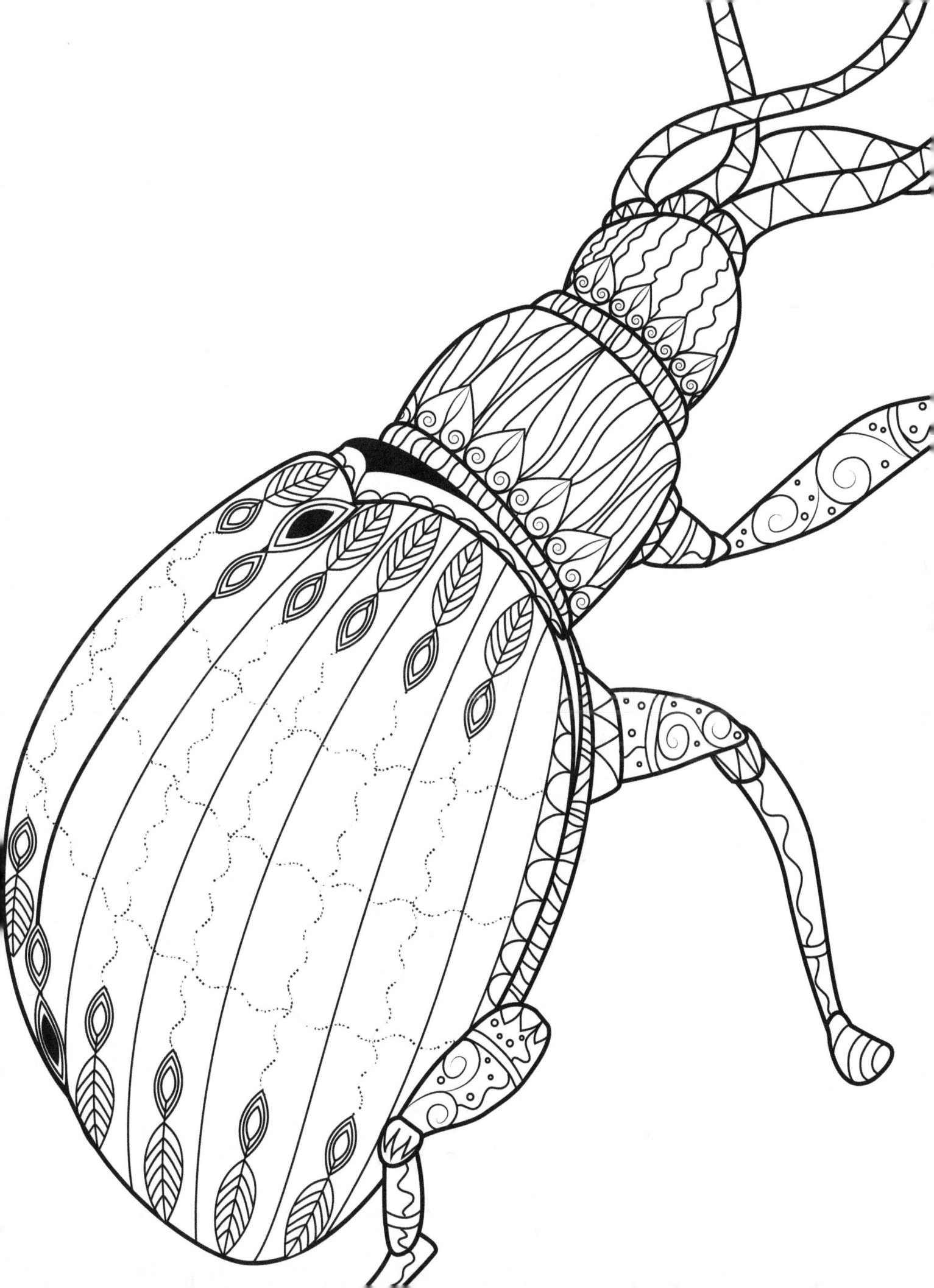

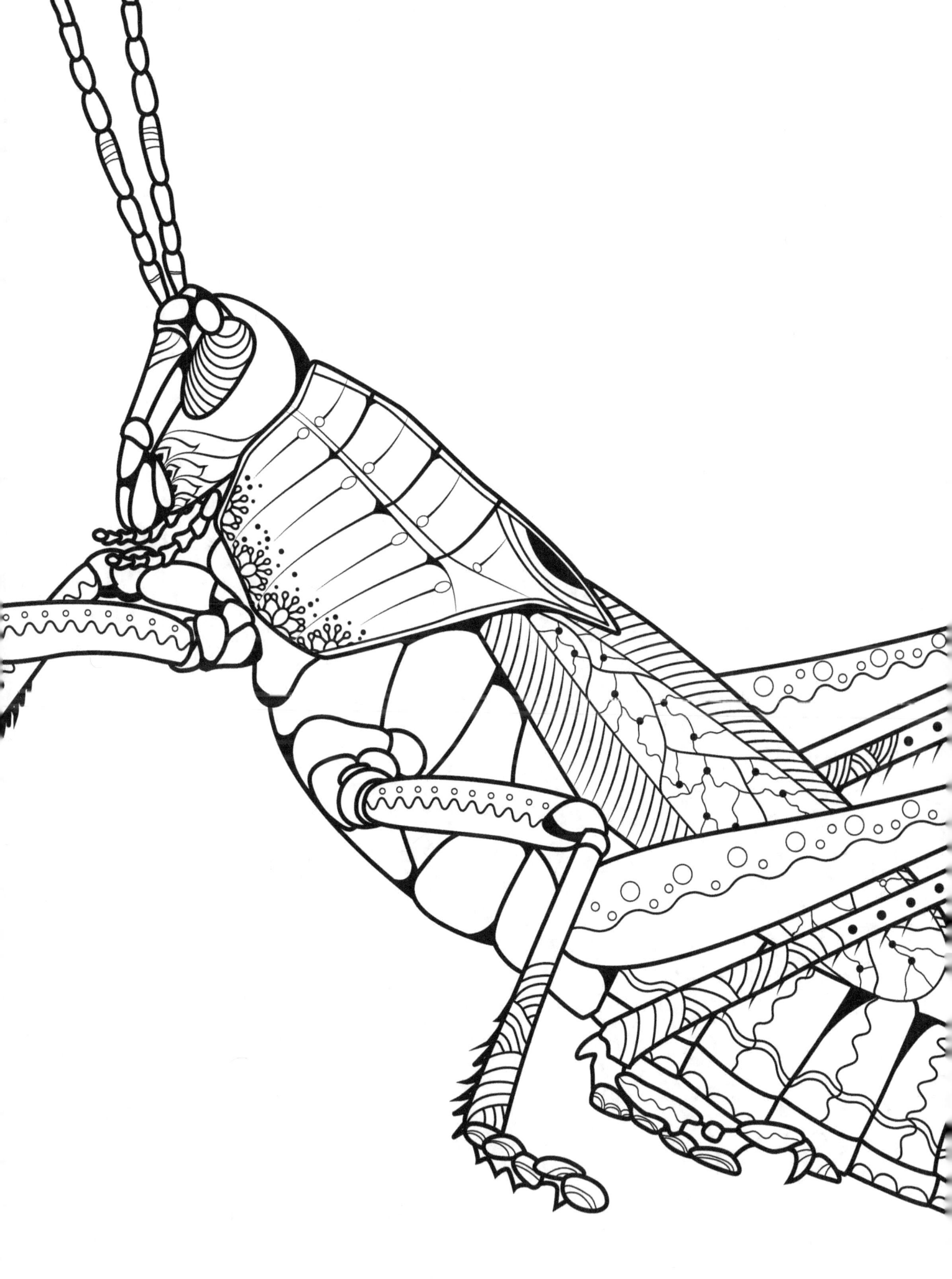

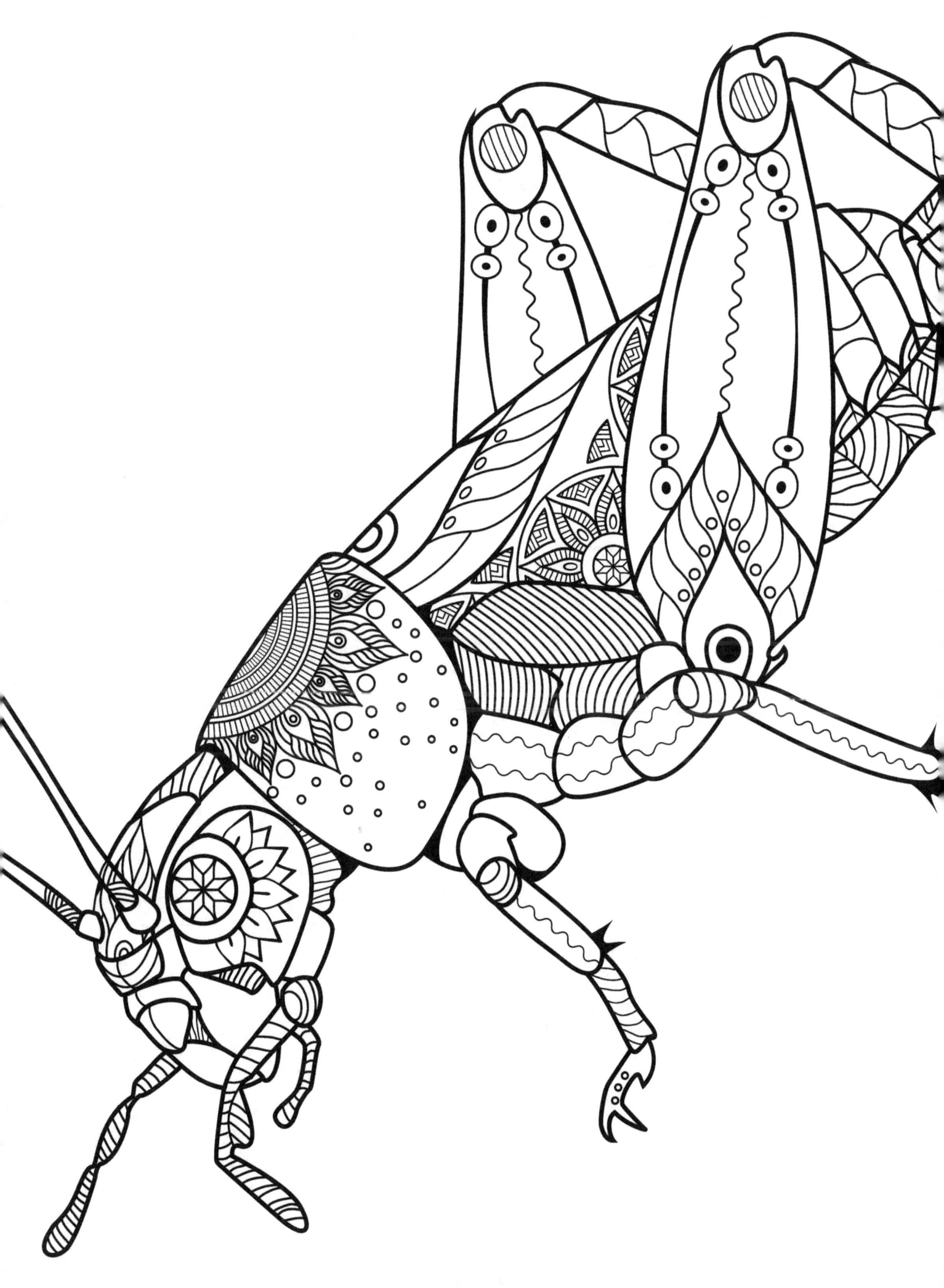

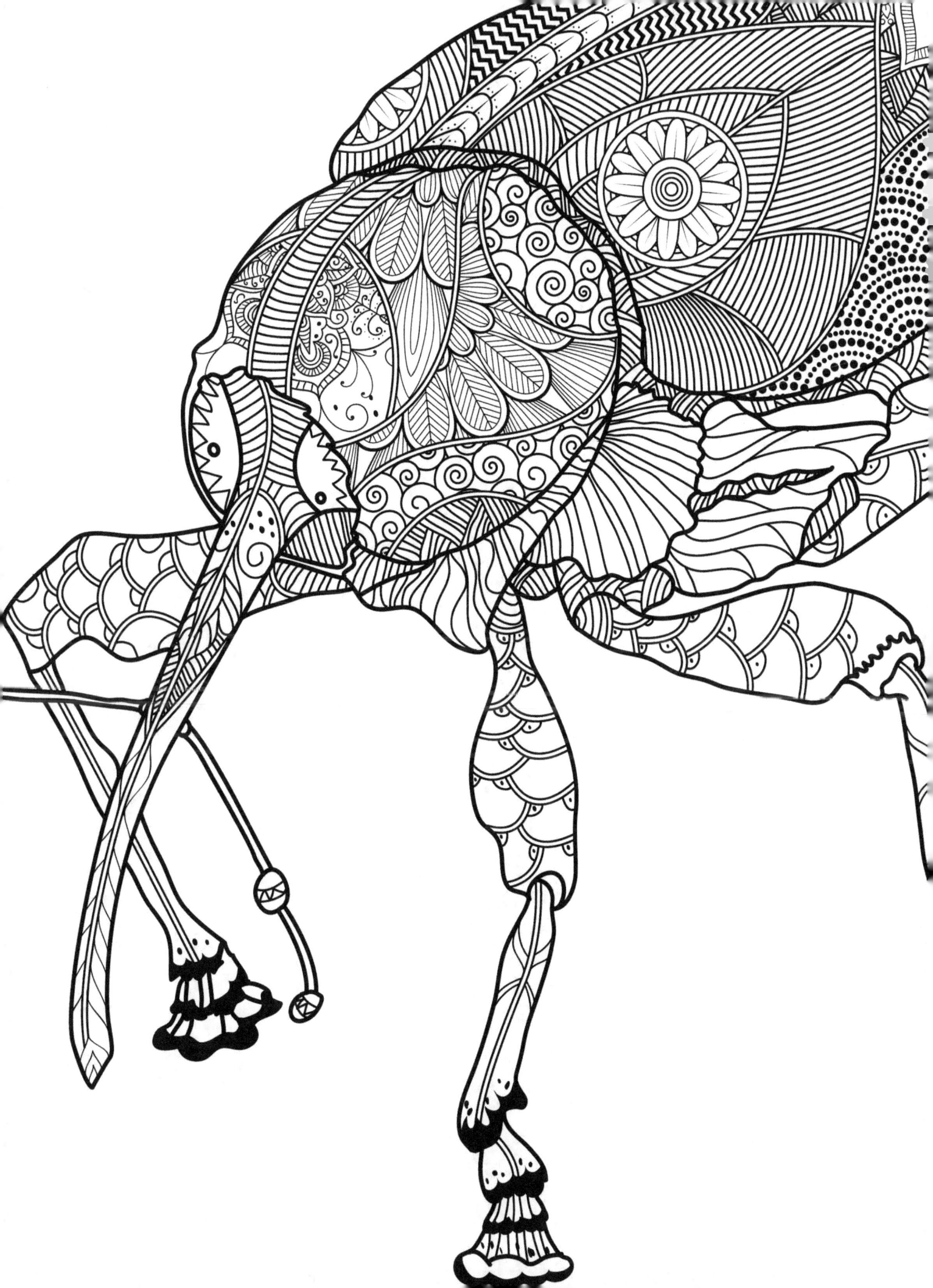

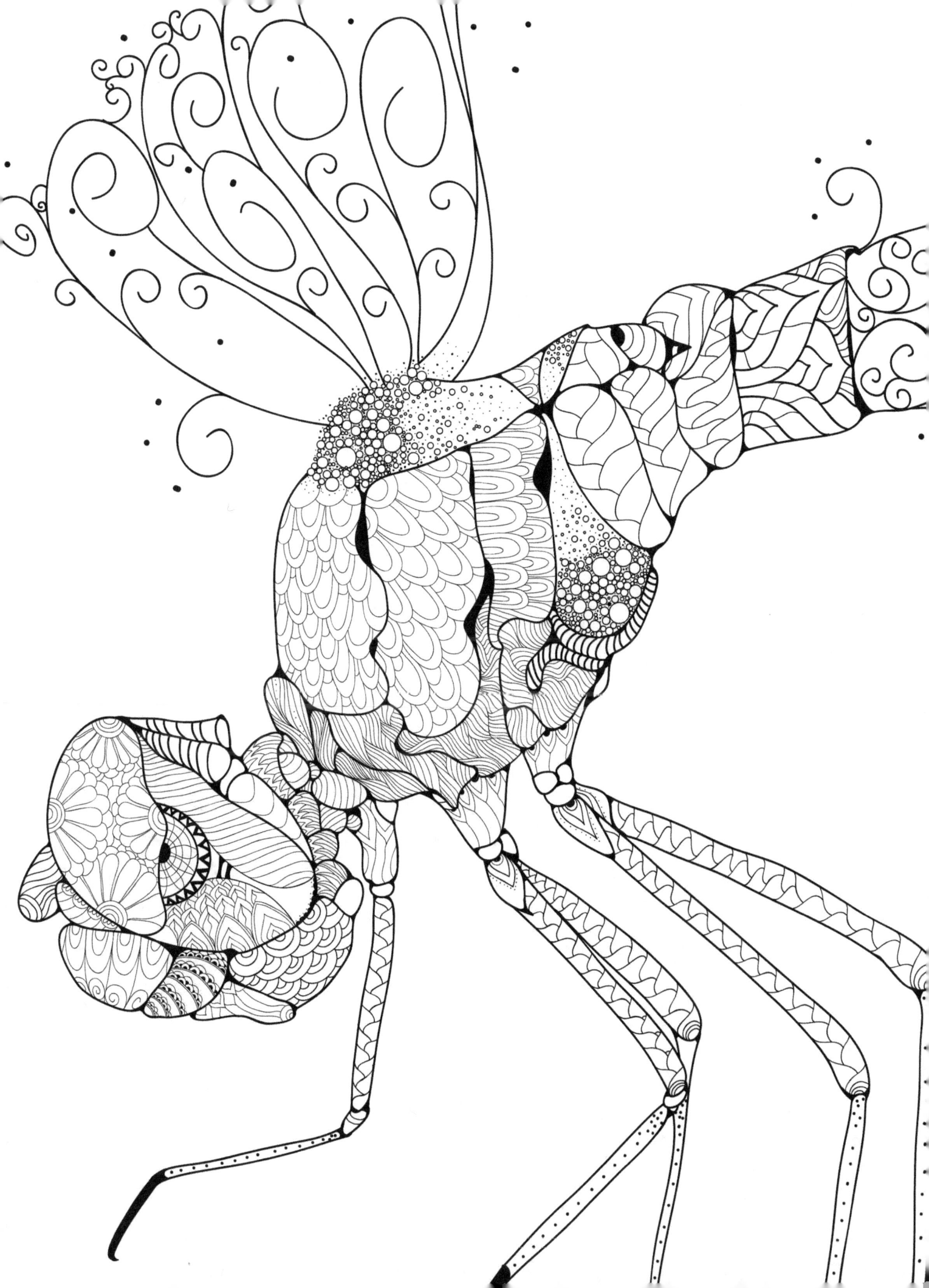

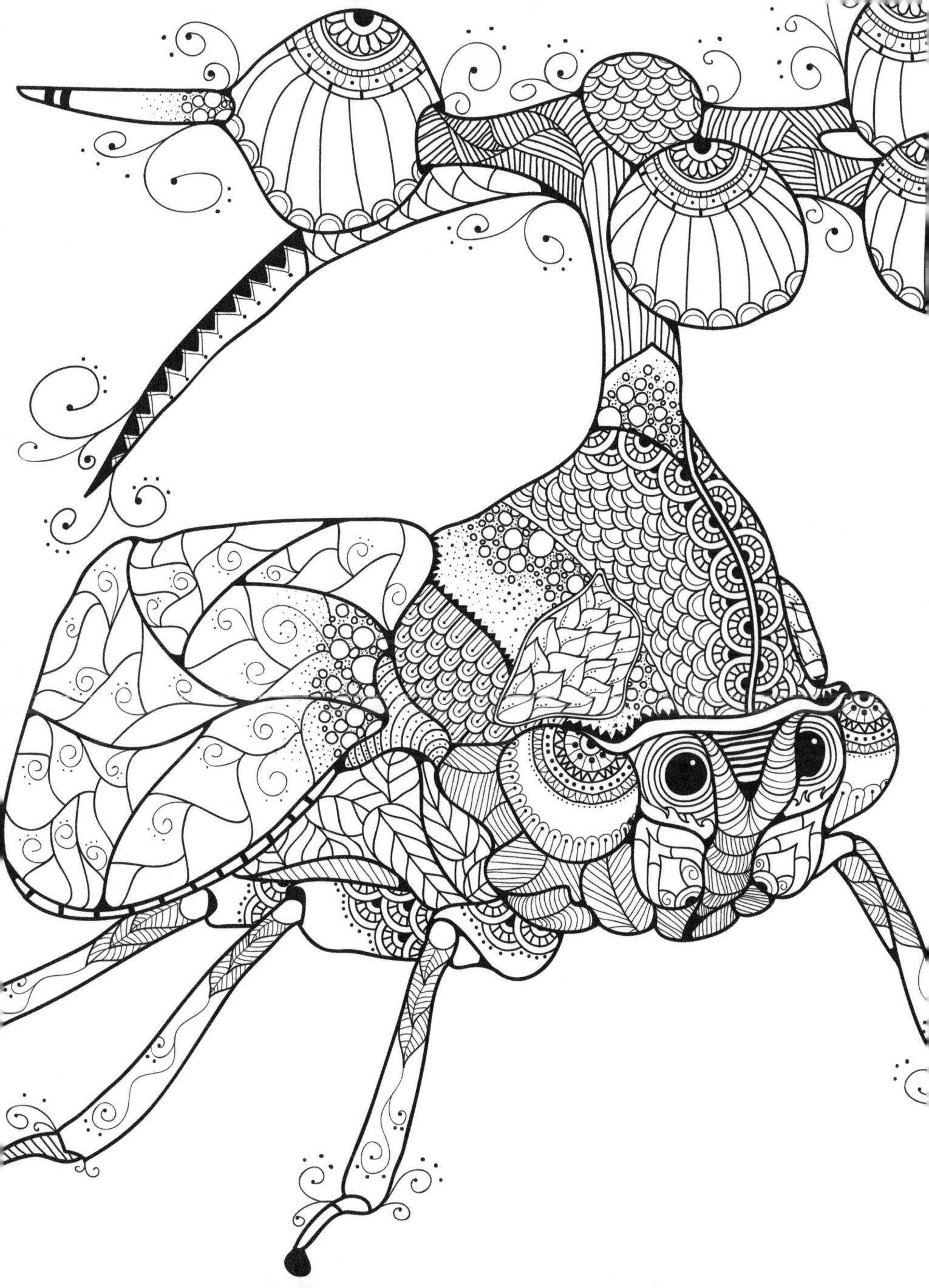

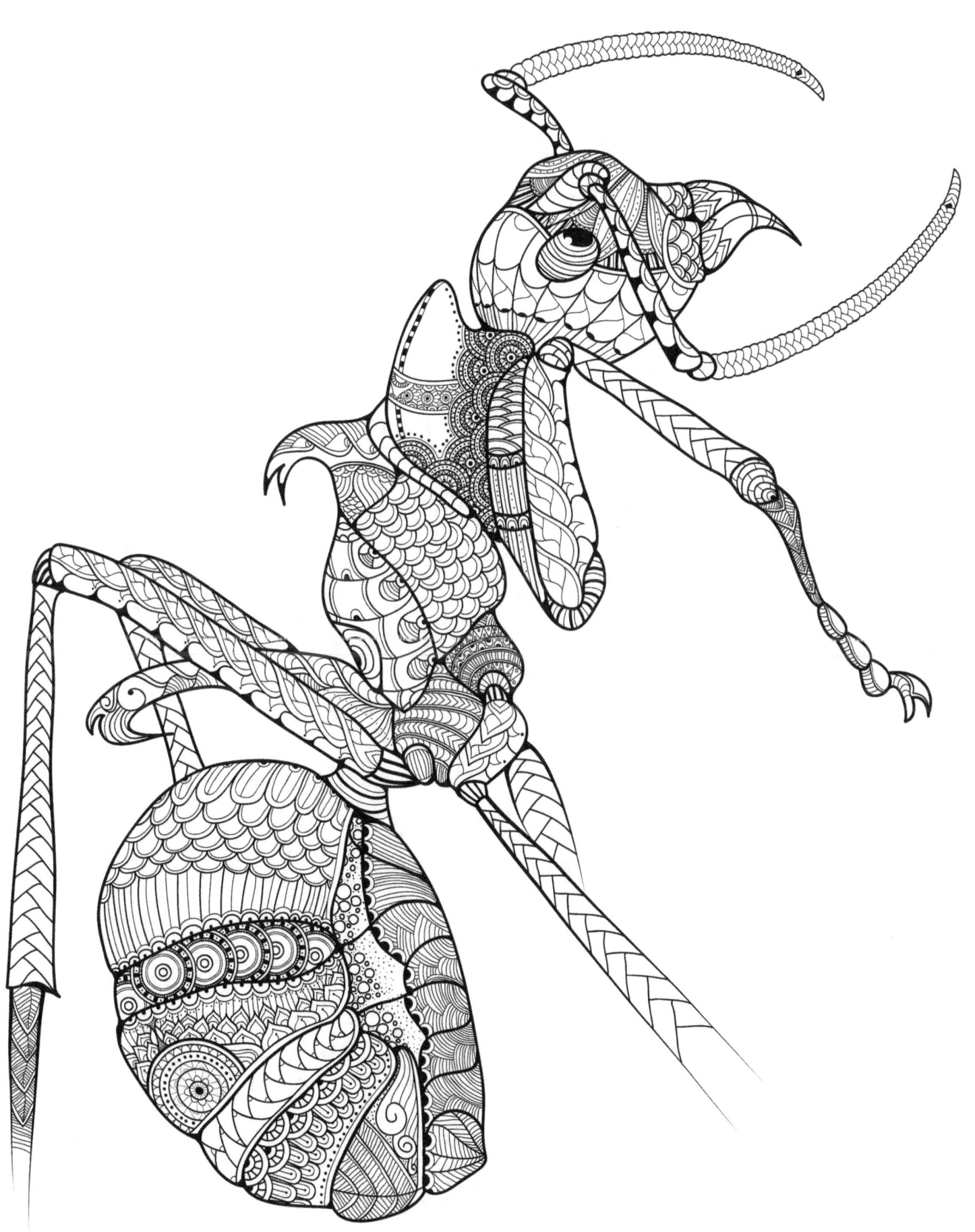

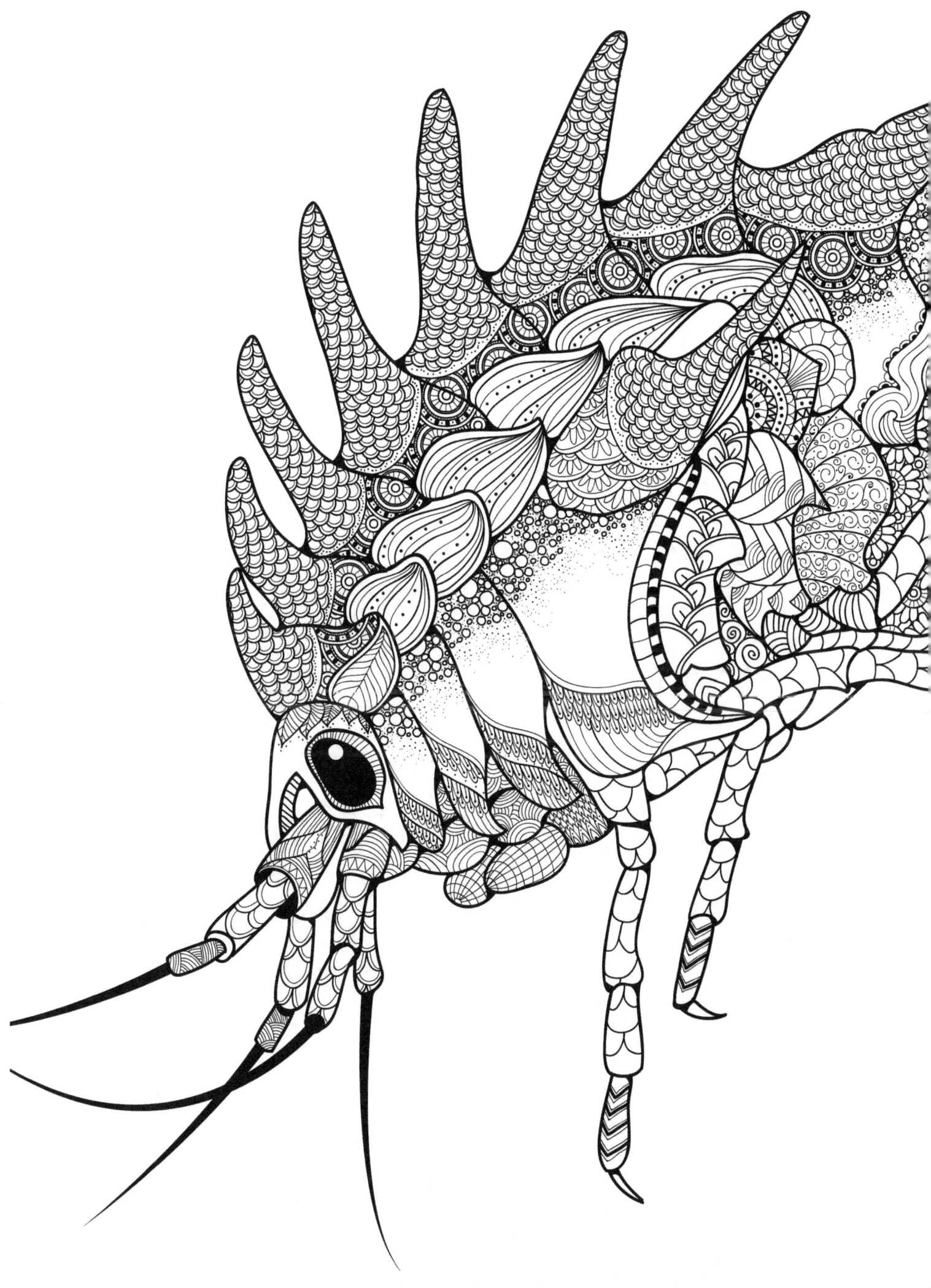

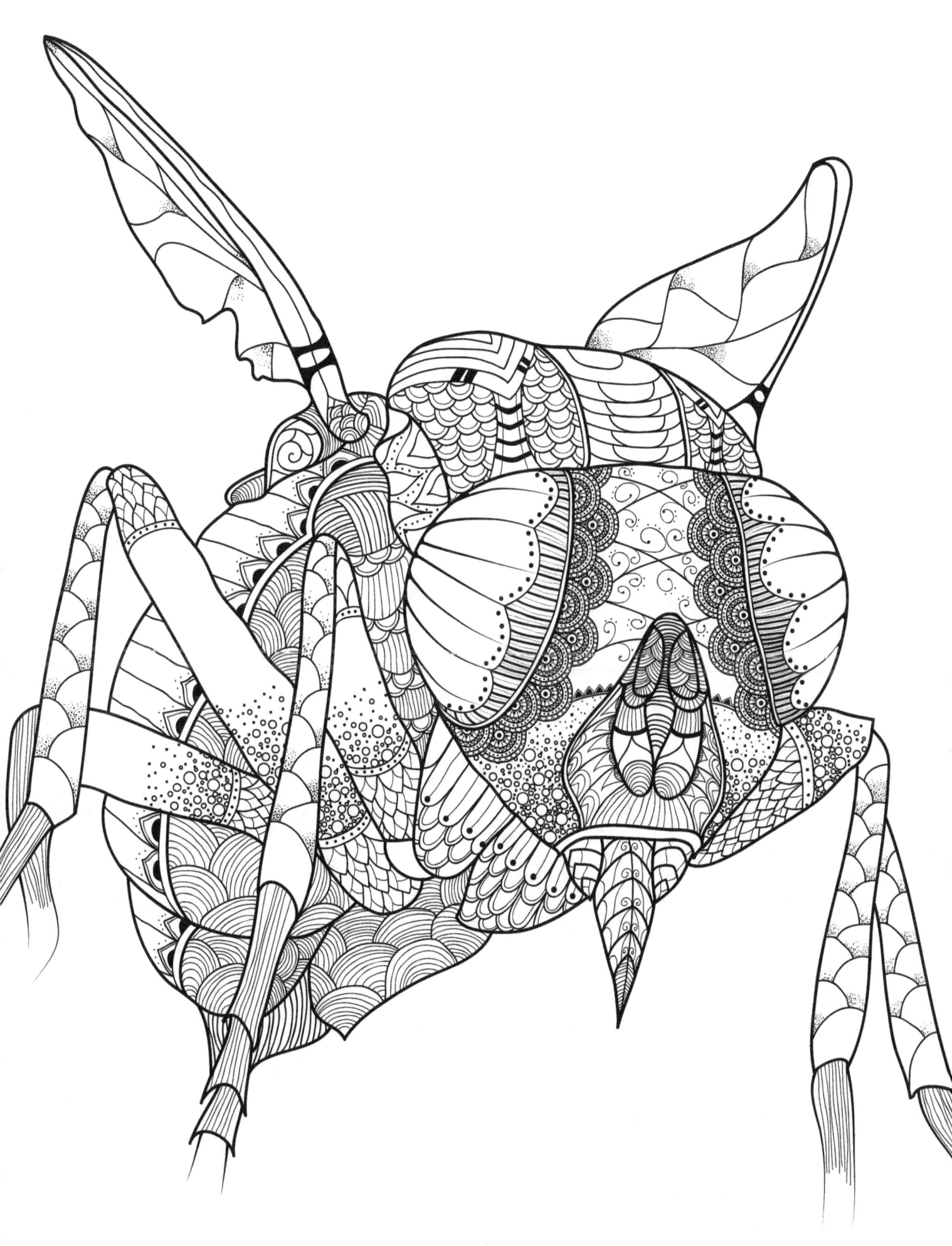

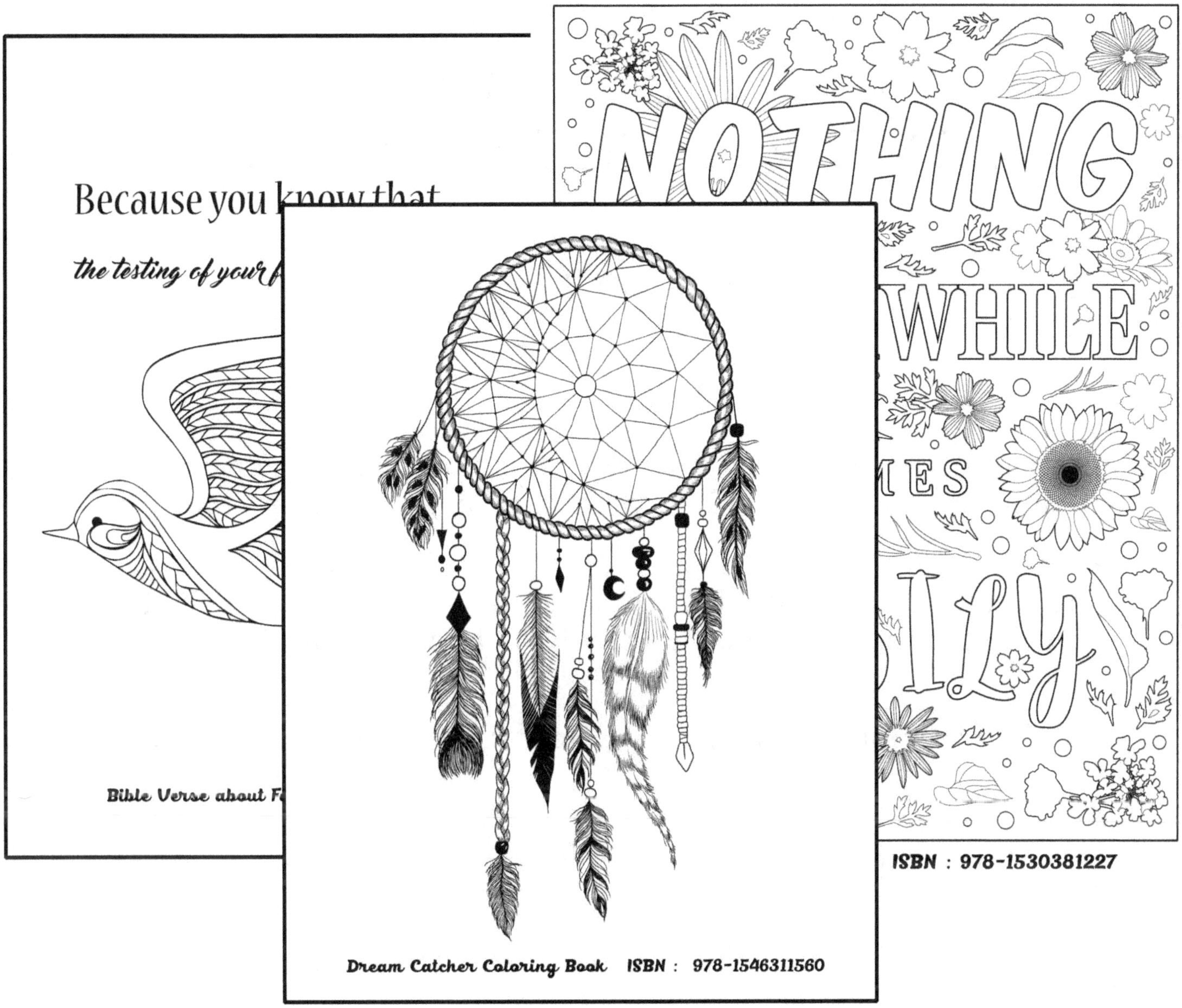

Free Download Coloring Pages

At : bit.ly/get_sample_free

Exclusive Offer
ONLY V Art Studio Fan Club!!

Join V Art Group : *http://bit.ly/join_cover*

- Coloring Challenge : Selected Works will be published in Our New Release and you also will get the commission of Sales (First 3 Months)

- Participate in Creating Our New Book : Book Cover Vote , Coloring Idea and more.. / Your name will be appear in our book.

- GET Free Coloring Pages/ New Coloring Books for Our New Release.

- Much More ...

Join Us at : *http://bit.ly/join_cover*

www.ingramcontent.com/pod-product-compliance
Lightning Source LLC
Chambersburg PA
CBHW082220220526
45470CB00010B/3248